Industrial Wastewater Management and Water Reuse Topical Conference

Topical Conference at the 2022 AIChE Spring Meeting and 18th Global Congress on Process Safety

San Antonio, Texas, USA
10 – 14 April 2022

ISBN: 978-1-7138-6814-9

Printed from e-media with permission by:

Curran Associates, Inc.
57 Morehouse Lane
Red Hook, NY 12571

AIChE
120 Wall Street, FL 23
New York, NY 10005-4020

Phone: (800) 242-4363
Fax: (203) 775-5177

www.aiche.org

Additional copies of this publication are available from:

Curran Associates, Inc.
57 Morehouse Lane
Red Hook, NY 12571 USA
Phone: 845-758-0400
Fax: 845-758-2633
Email: curran@proceedings.com
Web: www.proceedings.com

TABLE OF CONTENTS

641676 Sponsored Technology Workshop - Thermoplastic Design Considerations; Choosing the Ideal Material for a CPI Piping System - IPEX USA

Abstract:

This session will introduce different types of thermoplastics and discuss their unique and differentiating qualities. It will expand on relevant physical properties which allow thermoplastics to outperform traditional piping materials. Discussions will cover important design considerations which should be considered when selecting a material for an application. The workshop will also address common questions about the use and benefit of thermoplastics in industrial settings.

640413 Technology Innovations in Water and Wastewater Treatment Plenary Session

Abstract:

This plenary session consists of presentations from world experts in technology innovation in water. The speakers will give different perspectives on the drivers and barriers in moving new ideas from research into practice in the water space. The talks will include examples of new and innovative technologies for wastewater treatment, water treatment and water reuse.

639391 Safecrit and Digital Twin: An Integrated Set of Technologies for Protecting Water Treatment and Distribution Plants Against Extreme Cyber Attacks

Abstract:

In an extreme cyber attack the attacker sends commands to actuators and spoof sensor values in manufacturing plants. Such attacks aim to damage the plant, cause service disruption, and lead to unsafe or poor product quality. Several water plants have been the target of such attacks. The earliest known such attack is the SCADA security incident at Maroochy Water Services in Queensland, Australia, in 2000. The attacker took control of the sewage pumping stations that led to sewage flowing into local highways. A more recent attack occurred at the Oldsmar water distribution system in February 2021 when the attacker altered the settings of Sodium Hydroxide dosing to dangerous levels. Advisories from government agencies, and announcements from the White House indicate an increase in such attempts on water systems. This paper presents SafeCrit, an integrated collection of technologies, for protecting water plants against extreme cyber attacks. SafeCrit is embedded in a physical water treatment plant as well as in a digital twin described here.

SafeCrit technologies

A brief description of key technologies that comprise SafeCrit follows. DAD is a design-centric process anomaly detector. State entanglement, a patented method, is a key to the ultra-high anomaly detection rate of DAD. The idea is to use fundamental laws of physics to create relations across plant components. These relations, known as invariants, are programmed to monitor the underlying plant process.

AICrit uses machine learning and design knowledge to create models that capture the normal spatio-temporal relationships among sets of correlated components. It models the normal behavior of continuous-valued state variables (sensors) through the temporal dependencies to forecast their behavior. In addition, it also models the higher-order and non-linear correlations among the discrete (actuators) and continuous state variables.

SuperDetector fuses the inputs from real-time operational data from the plant with that received from AICrit and DAD. It uses a stacked ensemble of Random Forest, Bagging with k-Nearest Neighbours and Gaussian Na\"ive Bayes algorithms to classify normal and anomalous data. This results in high accuracy of detection and a low rate of false alarms.

PlantProtect thwarts an attacker's attempt at moving the plant to an anomalous state. It prevents a rogue command from the controller to reach its target actuator. A command sent to an actuator is first trapped and validated and released only when the actual and predicted commands are the same, else it the plant operator is requested for an appropriate action.

Embedding SafeCrit in a digital twin

iTrust has developed a digital twin that mimics the operation of a 6-stage water treatment plant named SWaT. The twin is a fully distributed system that includes over 50~processes that mimic the behavior of SCADA, Engineering Workstation, PLCs, RIOs, actuators, sensors, and the historian. A professional HMI, also developed at iTrust, is integrated with the twin. The twin uses the now industry standard OPC protocol for communications among PLCs and SCADA and the well known ZMQ protocol for communications across other processes. Modbus and ENIP protocols are currently being added to the twin as options.

All SafeCrit technologies have been integrated into the twin. Doing so has enabled researchers and educators to understand and test SafeCrit within a virtual environment at a much faster pace than they could on a live plant. The twin has been used in the world's largest cyber-exercise, named Locked Shields, conducted each year by the NATO Cooperative Cyber Defence Centre of Excellence. In this exercise, 26 instances of the twin were deployed and attacked by teams from NATO countries.

Quality requirements for SafeCrit

A major challenge faced by developers of cyber security technologies arises from the fact that extreme cyber attacks are rare when compared with the almost daily attempted, and often successful, network intrusions. However, when an extreme cyber attack occurs in a critical infrastructure, it could lead to significant financial and social chaos. This observation leads to stringent quality requirements for any cyber security technology aimed at defending water treatment and distribution plants.

From the onset, SafeCrit was required to demonstrate ultra-high rate of anomaly detection and ultra-low rate of false alarms. An ideal requirement is Zero False Alarms, i.e., ZeFA and 100% anomaly detection rate. While both requirements are equally important, ZeFA is often overestimated. Doing so may lead to a false alarm rate that might appear exceptionally low but may lead to an unusable technology. This observation arises when the rate at which data is received is high, e.g., every second. In this situation, even a 0.1% rate of false alarms would imply about 86 false alarms per day, which is obviously unacceptable for any plant operator. Thus, for SafeCrit, the rate of false alarms was set to 1 in 30~days and an anomaly detection rate of 95%.

Test infrastructure

In addition to the digital twin, iTrust offers two operational testbeds for water treatment and water distribution.These testbeds are referred to as SWaT for water treatment and WADI for water distribution. The testbeds mimic water treatment and distribution processes found in city-scale plants. These testbeds are used for continuous testing of SafeCrit technologies.

The Secure Water Treatment (SWaT) testbed contains a 6-stage water treatment plant producing 5-gallons per minute of water filtered using ultrafiltration and reverse osmosis. The Water Distribution (WADI) plant contains three stages to distribute 5000Litres of water to 6-consumers. The testbeds are connected to allow filtered water from SWaT to enter the primary grid in WADI, dosed chemically, and passed to the secondary grid. Water flows by gravity from the secondary grid to the consumer tanks. Booster pumps are used when the consumer demand cannot be met with gravity flow.

Testing SafeCrit

SafeCrit has been subjected to three increasingly intense levels of testing. At the lowest level, researchers involved in the development of SafeCrit tested it against an iTrust benchmark of extreme cyber attacks. This benchmark contains single-point and multi-point attacks that aim at deceiving a Programmable Logic Controller (PLC) into concluding that the spoofed data is correct. At the next level, SafeCrit is tested by independent attackers participating in an international exercise known as Critical Infrastructure Showdown (CISS) organized by iTrust. At the highest level, parts of SafeCrit are tested at city-scale water treatment plant.

Deploying SafeCrit

The deployment strategy for SafeCrit depends on the plant architecture. Deployment could be fully non-invasive or invasive. Non-invasive deployment is done by interfacing SafeCrit with the plant historian. Data from the historian is passed to SafeCrit via a data diode ensuring one-way communication and plant safety. Thus, at every sampling instant, SafeCrit receives plant state via the historian and generates an alert when the plant moves into an anomalous state.

Non-invasive deployment limits SafeCrit's capability in realizing its objectives. For example, in non-invasive mode, PlantProtect is unable to receive data from level~0 network and hence cannot use the trap-validate-release strategy. However, it can still correct sensor states based on predicted values. While state predictions in SafeCrit are accurate, a clever attacker can still deceive SafeCrit by launching multi-point attacks.

Invasive deployment requires creation of the level~0 network. This is done by tapping the commands from PLC sent to the actuators via the Remote IO (RIO). In addition, selected invariants in DAD are also added to the control code in each PLC. Such deployment leads to a significantly more robust SafeCrit making it extremely challenging for an attacker to bypass its detection and protection mechanisms.

Can SafeCrit fail?

SafeCrit's anomaly detection rate is high but not 100%. Its detectors may fail, for example, when an attacker is aware of the thresholds used. These thresholds are carefully tuned to lead to ZeFA and 100% detection. A ``slow attack" that remains below the thresholds, may remain undetected for long. However, experiments have revealed that such attacks might fail to cause any plant damage or service disruption given the standard engineering practice of using safety margins during design. For example, the overflow level of a water tank is usually set to below the true height of the tank. Thus, while a ``virtual overflow" could occur, an attack would be detected before water actually overflows the tank. While SafeCrit has been able to achieve ZeFA over long plant runs, it does not, and cannot, guarantee such performance at all times. Thus, alarms could begin to occur when plant components degrade requiring retuning of SafeCrit parameters.

Status of SafeCrit

SafeCrit technologies are currently being licensed for conducting pilots and incremental deployment. (Contact: Director, Office of Research and Industry Collaborations; research@sutd.edu.sg).

639394 Impact of Cool Water Blowdown from District Cooling on Surface Water Using Water Quality Analysis Simulation Program (WASP 8)

Abstract:

Cooling towers are primarily used for heating, air conditioning, ventilation, and other industrial purposes. They are cost-effective and far more efficient than alternative dry systems. However, due to the water loss caused by evaporation, leakage, etc. the concentration of ions increase, which could lead to corrosion and scale formation (e.g. magnesium, calcium, silica, scale inhibitors, and chloride) remain inside the tower. This can have detrimental effects on the tower and lead to corrosion problems. To prevent such operational issues, a relatively great volume of cooling water needs to be discharged regularly as blowdown and replaced with fresh water as make-up to the tower. Despite being a viable solution, the discharge of cooling water blowdown (CWBD) into water bodies must be studied to determine if there are any adverse impact on marine life. In the state of Qatar, district cooling facilities play a major role in providing large-scale cooling needs to business districts, schools, etc. There is a need to find a suitable and practical system for the management of the cool water blowdown discharge. One alternative is to consider discharge to surface water. Doha Bay which is located on the Arabian Gulf, bordering Doha, the capital city of the State of Qatar. Due to water scarcity, Qatar has certain regulations and permissible limits for industrial discharges into surface water outfalls to prevent any negative impacts on the sea. Hence, it is important to conduct a sea outfall study to investigate the fate and transport of the contaminants in the CWBD stream using a water quality modelling tool and to assess the long-term effect of such contaminants on the marine life.

Studies related to sea outfall in the Arabian Gulf are scarce, which mainly aims to investigate the ambient environment of seawater. However, risk assessment of discharging contaminants associated with CWBD in Qatar requires more research. Therefore, this study aims to examine the fate and transport of contaminants associated with CWBD in Doha Bay using the Water Quality Analysis Simulation Program (WASP8). The EPA's WASP8 is a model that provides the users with the option to interpret and predict water quality responses to natural phenomena and manmade pollution. Two main outfalls in the Bay area are selected to model the daily variations of CWBD constituents in both outfalls. Inputs to the model, such as outfall segmentations, bathymetric data for each segment, a volumetric flow rate of CWBD discharge, inflow concentration of constituents, kinetics, and weather data will be used to develop a WASP8 model. Seawater samples are taken at each segment for both outfalls and are analyzed in the lab to determine the concentration of the main constituents of CWBD and their kinetic processes that govern the fate of the CWBD contaminants. Finally, a sensitivity analysis will be applied to investigate both, the impact of varying the volumetric flow rate and dilution effect on the fate and transport of the modelled variables

639438 Using a Wastewater Process Model to Evaluate the Effect of Mixing Cooling Water Blowdown (CWBD) on the Performance of WWTP: A Case Study

Abstract:

District cooling systems in Gulf region are generating huge volumes of cooling water blowdown (CWBD) that is typically rich in chemicals and contaminants. Cooling water blowdown was previously dumped straight into surface water bodies and never been treated or reused. Discharging CWBD blowdown water can cause several environmental problems. Due to the dangerous nature of the disposal options, other sustainable options are explored. The kind and quantity of contaminants present in CWBD differ from one system to other and influenced by a variety of variables. For example, the source of the cooling tower's input and make-up water has a significant influence on the presence of various pollutants. Chemicals employed for treatment, as inhibitors, or even as anti-corrosion inside the towers, can have a significant impact on the chemical makeup of the water. Some contaminants are typically seen in CWBD samples such as high TDS, high metal content, low biodegradability, and the presence of chemicals such as calcium, sulfates, silica, phosphates, are some of the most common concerning contaminants in the study of CWBD characteristics. This paper will study the possibility of discharging CWBD into already functioning wastewater treatment plants rather than building a treatment plant exclusively for blowdown. The effect of CWBD on wastewater treatment plants that employ a biological reactor will be explored. Doha North Wastewater Treatment plant (DNWWTP) was considered as the case study and the plant was simulated using GPS-X for theoretical results. The main concern is that the contaminants present in CWBD will have an adverse effect on the biological reactor. Mixing different percentages of CWBD with the wastewater influent in the GPS-X model showed no adverse effects on the performance of the biological reactor nor the overall systems removal efficiency of the plant. 5%, 10% and 20% v/v CWBD was taken into consideration for the study. The biodegradability, effect of adding CWBD in the influent if WWTP was performed using BOD tests & organic matter measurements for 5 days. The toxicity and inhibition effect of CWBD were measured in 5 identical ASR each mixed with different % of CWBD The lab-scale results, biodegradability, toxicity and inhibition tests as well as GPS-X results conclude that mixing CWBD up to 20 v/v% ratio with the WWTP effluent has no effect on the performance of the process nor develop inhibition/ toxicity or salt accumulation.

639642 Development of Optimization Model and Evaluation for Biological Wastewater Treatment Process

Abstract:

Efficiency determination of the aeration tank for biological wastewater treatment requires the use of computer simulation and laboratory study. The main purpose of the presentation is to talk about development of optimization model and evaluation for the operation efficiency of aeration tanks. An optimization model has been developed for the computer calculation of the biological wastewater treatment process in aerotanks. The model is based on mass conservation equations and the Monod model. The software implementation of the constructed model has been carried out. The results of a computational experiment on the study of the wastewater treatment process in an aeration tank are presented. This leads to the possibility for the quality control of wastewater treatment in aeration tanks.

639278 The Development of an Ionic Liquid-Based Demulsifier for Oil/Water Separation

Abstract:

The petroleum industry produces a huge quantity of wastewater in the form of emulsions that contains a significant fraction of oil. The separation of oil from the produced water is a significant challenge that is being addressed using different types of techniques. Chemical demulsification is one of the most important The application of ionic liquids (ILs) as a demulsifier is considered as an emerging technology in the petroleum industry. The primary objective of this work was to evaluate the performance of the ionic liquids for the demulsification of water in oil (W/O) emulsions at different conditions such as ILs dosage, structure (anions, and alkyl chain), and salt concentration. A bottle test was used to measure demulsification efficiency (DE%). Firstly, a crude oil emulsion was prepared with deionized water and oil at a different ratio. The demulsifier was prepared by dissolving ionic liquids in a solvent at specific concentrations. Emulsion characterization was carried out with the average droplet size measurement analysis and rheological analysis. The results from the current experimental work with pyridinium-based ILs with different anions indicate that a higher degree of hydrophobicity improves demulsification. A high temperature of 90°C was used to accelerate the demulsification of W/O crude oil emulsion. Demulsification efficiency of above >95% was obtained with more hydrophobic anions in no time at a concentration as low as 500 ppm. When it comes to structural configurations, the demulsification of the W/O crude oil emulsion was favored by the use of more hydrophobic IL's containing butyl alkyl chains. The least efficient demulsifier was found to be the one with the shortest alkyl length and the least hydrophobic anion. The involvement of seawater accelerated the separation with all the ionic liquids resulting in high demulsification efficacy of the emulsion. The application of the pyridium class of IL as a demulsifier for the demulsification of crude oil is very rare and scarcely available. This study helps us to carry out a comprehensive analysis of evaluating the available ionic liquid and their ability to carry out efficient separation of water in crude oil emulsions.

640208 Polymer Nanofiber Mats for Selective Removal of Pfas from Wastewater

Abstract:

This paper will present a novel technology using nanofiber mats fabricated by electrospun polymer nanofibers for selective removal of per and polyfluoroalkyl substances (PFAS) from landfill leachate. The co-contaminants, low concentration, and complexity for PFAS in industrial wastewater are big hurdles for current technologies. This method overcomes the challenges and provides a unique approach for PFAS removal. The method is a hybrid and synergistic PFAS removal with adsorption/membrane, with smart and selective adsorption feature to reduce PFAS concentration to below 70 ppt, and capable of handling co-contaminants and high loading flow rate.

The polymer nanofiber mats are regenerable after PFAS recovery for further destruction. This is a cost-effective and energy-efficient approach.

638512 Graphene Oxide Quantum Dots Coupled with Advanced Oxidation Processes for the Removal of Pharmaceuticals and Disinfection of Wastewater

Abstract:

Graphene oxide quantum dots (GQDs) are an interesting class of zero-dimensional, oxygen-rich materials with size less than 20 nm. GQDs exhibit the phenomena of quantum confinement and edge effects which bestow them with outstanding physico-chemical properties. These features have made GQDs suitable materials for various applications including the catalytic removal of organic pollutants and the disinfection of microbes. On the other hand, peracetic acid (PAA) is a strong disinfectant with a wide spectrum of antimicrobial activity. Due to its bactericidal, virucidal, fungicidal, and sporicidal effectiveness as demonstrated in various industries, the use of PAA as a disinfectant for wastewater effluents has attracted much recent attention. PAA is referred to as green disinfectant widely used in wastewater treatment.

This study aimed to investigate the synergy advantage of combining GQDs and PAA based advanced oxidation processes (AOPs) for the removal of pharmaceuticals and disinfection of wastewater at tertiary stage of the treatment plant.

A series of experiments were carried out: a) degradation efficiency and kinetics of GQDs/PAA system against pharmaceuticals in real wastewater; b) identification of the main reactive radicals responsible for the degradation of pharmaceuticals, c) photodegradation mechanistic pathway based on the detected reaction by-products determined by LC-Q-TOF-MS; d) antimicrobial activity; e) Genotoxicity and mutagenicity test.

The results indicated that the same degradation trends of pharmaceuticals were observed for the influence of PAA dosage; influence of water matrix and influence of initial concentration as a function of contact time.

Furthermore, the main reactive radicals responsible for the degradation of pharmaceuticals in wastewater were identified as hydroxy (•OH) as well as peroxy radicals CH3C(=O)OO• and CH3C(=O)O•. The genotoxic and mutagenic potential of the degradation products formed during the degradation of pharmaceuticals were confirmed to be non-mutagenic. GQDs/PAA was further tested as a potential disinfectant and *S. aureus* was completely inactivated. GQDs/PAA also eliminated more than 90% of bacteria present in real wastewater.

This contribution presents an opportunity to simultaneously degrade pharmaceuticals and their active metabolites as well as inactivate microorganisms using GQDs/PAA. These results are crucial and might find application in water/wastewater treatment plants.

639540 High-Strength Wastewater to Bioproducts through the Carboxylate Platform

Abstract:

In the circular economy, waste-to-energy (WTE) technologies are essential to minimize waste and pollution, reduce dependency on fossil fuels, and combat climate change. The carboxylate platform is a promising approach to sustainably produce valuable bio-derived chemicals and fuels from low- and negative-value waste streams. With the expansion of global economy, large volumes of high-strength wastewater (HSW) streams are discharged into the environment and cause severe problems (e.g., eutrophication). When treating HSW, traditional wastewater treatment processes are hindered by low treatment efficiency and high operation costs; thus, a new cost-effective HSW treatment process is needed.

A novel integrated HSW treatment process based on the carboxylate platform was developed to valorize HSW. The process enables simultaneous HSW treatment and carboxylate production via high-rate methane-arrested anaerobic digestion (MAAD). This study specifically determined the links between HSW characteristics, microbial consortia structure, and the design and operation of high-rate MAAD from bench-scale (500-mL) to pilot-scale (100-gallon) systems. The objectives of this study include: (1) establishment of a tailored microbial consortia for high-efficiency HSW treatment and high-titer carboxylate production, (2) development of MAAD using actual dairy and brewery wastewater as the example HSW, (3) separation of MAAD broth via *in-situ* and downstream operation modes, (4) development of MAAD kinetic model, and (5) techno-economic analysis to determine the optimal economic performance for specific bioproducts (e.g., sustainable aviation fuel).

Specifically, a robust microbial consortium with high-salt toxicity tolerance was selectively established to achieve high treatment efficiency and acid productivity. A blend stream of cheese whey and brewery wastewater at high COD concentration (>70 g/L) was treated by MAAD with established microbial consortium to produce carboxylic acids with high concentration (>40 g/L) and COD conversion (>80%) at bench- and pilot-scale digesters. This study provides a comprehensive investigation of the effects of process operating conditions and digester configurations (e.g., fed-batch digester, anaerobic membrane digester, electrodeionization-assisted anaerobic digester) on the HSW treatment and carboxylate production. Furthermore, to determine key microbial players, the microbial community structure and dynamics were also analyzed under high COD concentration and acidic environment. The results show that the newly developed process has the potential for large-scale application. This is a successful example of the carboxylate platform to transform low- or negative-value waste streams into high-value bioproducts.

Headworks INTERNATIONAL

Headworks International Inc.

Headworks INTERNATIONAL

Life Cycle of Water in Petroleum Industries from Exploration & Production to Refining and Downstream Processes

A Presentation at the Industrial Wastewater Management & Water Reuse Topical Conference:

AIChE Spring 2022 Meeting

April 12, 2022

Henry B. Gonzales Convention Center, San Antonio, TX

by

Somnath Basu

Headworks International, Houston, TX, USA

Outline

- Health and Safety Statement
- Upstream E & P and Water Treatment Overview
- Typical Petroleum Refining Process
- Total Water Network in a Petroleum Refinery
- Fire Water System
- Stormwater Handling and Disposal
- Water Treatment
- Process Water Management
- Handling and Management of Water Streams in a Refinery
- Fate of Treated Wastewater and Environmental Impacts of Discharge

Abbreviations

- AOC = Accidentally Oil Contaminated water
- BFW = Boiler Feedwater
- BOD = Biochemical Oxygen Demand
- CCU = Cat Cracking Unit
- COD = Chemical Oxygen Demand
- FS = Filterable Solids
- DM = Demineralized
- DO = Dissolved Oxygen
- EOF = Entirely Oil Free water
- ETP = Effluent Treatment Plant
- GAC = Granular Activated Carbon
- HAP = Hazardous Air Pollutants
- IX = Ion Exchange
- RO = Reverse Osmosis
- TOC = Total Organic Carbon
- LP = Low Pressure
- LTO = Light Tight Oil
- NPDES = National Pollutant Discharge Elimination System
- PAC = Powdered Activated Carbon
- SWS = Sour Water Stripper
- SSW = Stripped Sourwater
- TAN = Total Acid Number
- TMDL = Total Mass Based Daily Load
- TN = Total Nitrogen
- TSS = Total Suspended Solids
- UF = Ultrafiltration

Health & Safety Statement

- Operation, maintenance and working around refinery wastewater treatment plants may cause potential health and safety hazard without proper HSE program
- Types of hazard
  - Chemical: health risk from exposure to H_2S, VOC, ammonia, amines, acids, caustic soda, etc.
  - Biological: health risk from exposure to microbiological species
  - Reactivity and explosivity: due to trapped gases in the covered equipment headspace
  - Physical: injury from slip, trip, fall and vehicular accidents
- Refineries need to train staff and implement strict HSE program
- O&M personnel need proper safety gear to protect themselves against injury, long term health risk and death
- Safety is of utmost importance and cannot be compromised

Upstream Exploration & Production Operation and Produced Water Treatment

Typical Petroleum Upstream Facility

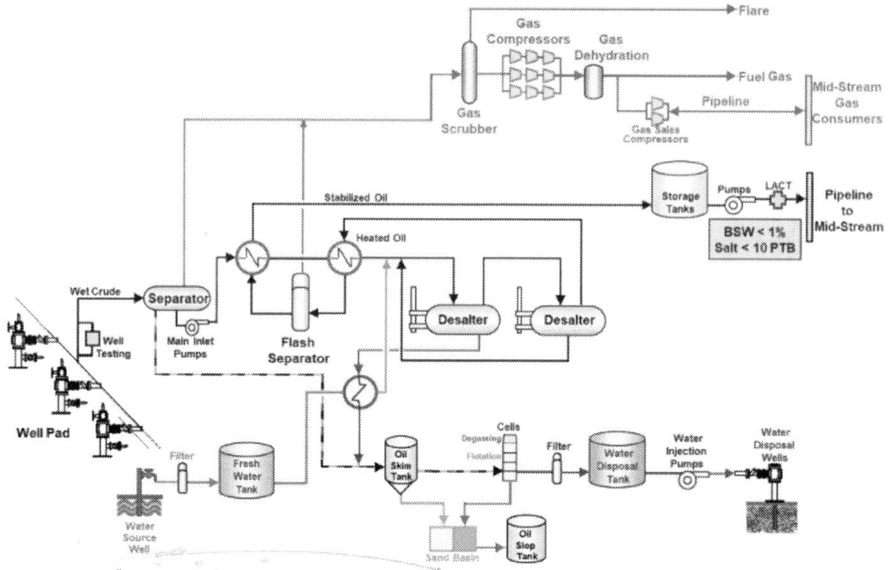

Typical Shale Oil Produced Water Treatment

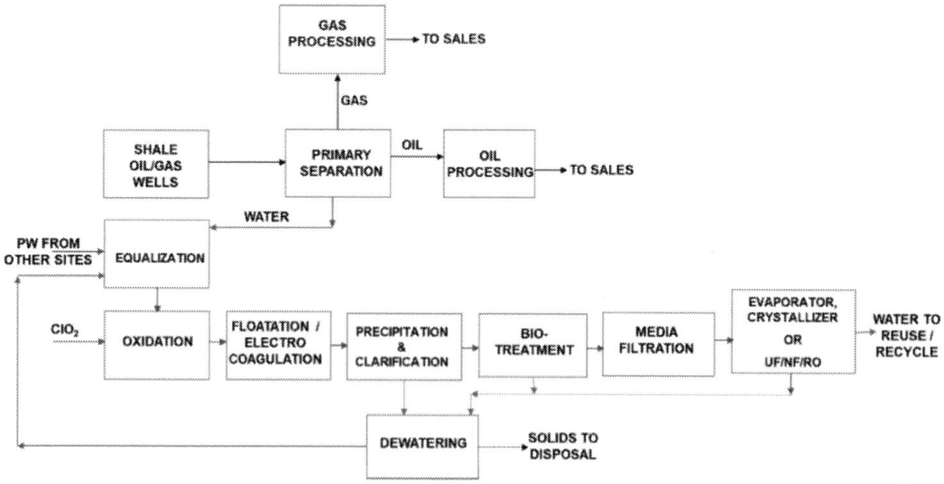

Petroleum Downstream Processing

Typical Petroleum Refining Process

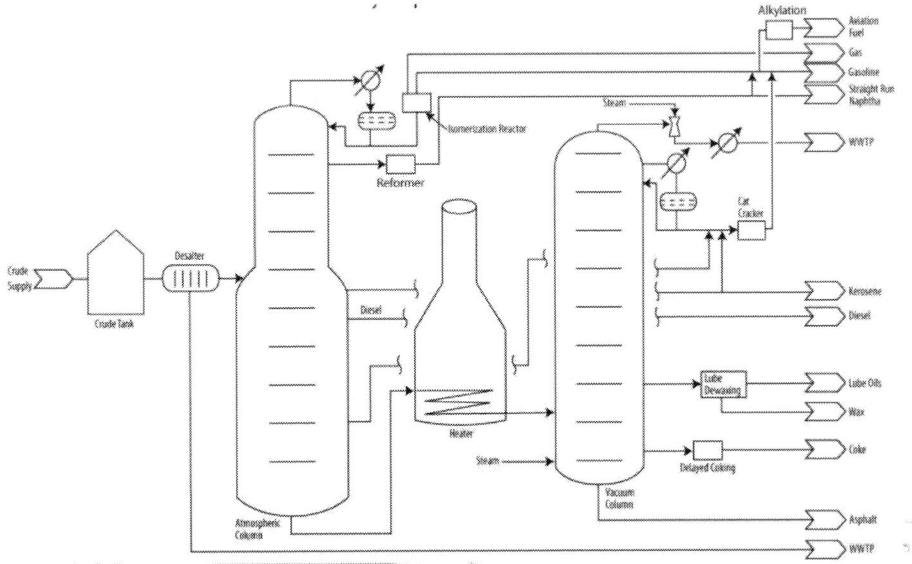

Total Water Network in a Petroleum Refinery

- Stormwater Handling and Disposal
  - Entirely Oil Free (EOF) water
  - Accidentally Oil Contaminated (AOC) water
- Tank Dike Water and Ballast Water
- Firewater Storage and Handling
- River/Sea Water
- Feed Water Treatment
  - Water treatment
  - Cooling water blowdown disposal
  - Demineralization
  - Condensate polishing
  - Boiler blowdown and demineralizer regeneration waste disposal
- Process Water Management
  - Sour water stripping (continuous)
  - Crude desalting and brine disposal (continuous)
  - Vessels steam out/clean out condensate streams (discontinuous)
  - Collected oily water from drips and drains (discontinuous)
- Domestic Sewage and Treatment
- Wastewater Treatment
  - Primary Treatment - Physicochemical
  - Secondary Treatment – Biological
  - Tertiary Treatment – Final Polishing

Headworks INTERNATIONAL

Petroleum Refinery Water Usage: 0.5 to 1.5 BBL of Water / BBL of Crude Processed

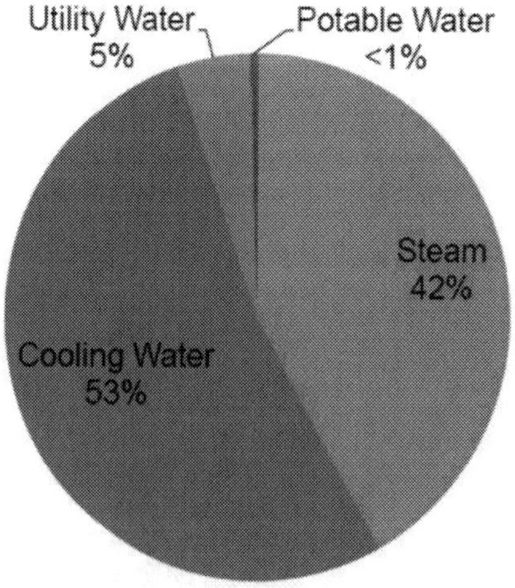

Total Water Network in a Petroleum Refinery

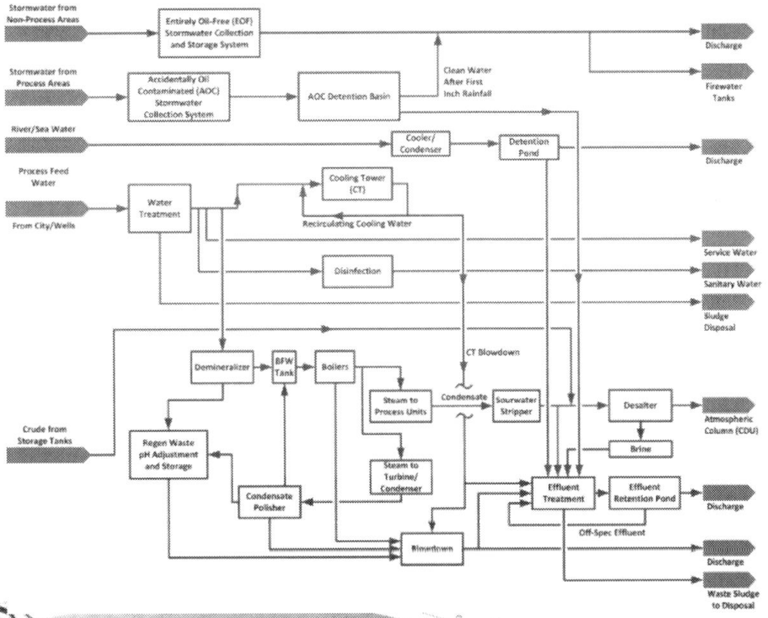

Fire Water Storage and Handling

- Firewater system very important in a refinery to combat fire and explosion incidents
- Typically consists of
  - A firewater pond for storage
  - Set of continuously running pumps
  - Water recirculating through a pressure piping system throughout the refinery with hydrants for tapping water as needed
- Various sources of fire water
  - Treated wastewater effluent from retention pond
  - Clean stormwater from detention basin
  - River or lake water

Stormwater Handling and Disposal

Headworks INTERNATIONAL

- Collection system of surface drains distributed throughout refinery
- Open channels lead to network of underground closed pipes
- Separate collection systems for EOF runoff and AOC water
- EOF runoff directly discharges to receiving water
- AOC stream led to a detention basin
  - First flush rainfall contained and tested, if found contaminated then slowly released to wastewater treatment plant in 48 hrs.
  - Water collected beyond first flush rainfall slowly discharged through EOF

Headworks INTERNATIONAL

River/Sea Cooling Water Systems

- Once through open loop system
  - No cooling tower
  - Only treatments:
    - Screening to remove large objects at the intake line
    - Disinfection by a biocide
  - Return stream discharged to receiving water body through detention basin to ensure no hydrocarbon contamination
  - Temperature increase may be a concern

Feed Water from City/Surface Resources/Wells

- Feed water sources
  - City water supply
  - Surface water from rivers, or lakes
  - Groundwater from wells
- City water may not require any treatment prior to demineralization for boiler feed
  - May need treatment by activated carbon (GAC) to remove residual chlorine and TOC (humic and fulvic acids)
- River/lake water needs treatment to remove hardness, suspended solids, turbidity and TOC
- Groundwater needs treatment to remove dissolved iron, manganese, hardness and TOC

Water Treatment

- Water treatment makes supplied water suitable to feed demineralizers, sanitary water and service water loops
- Dissolved iron and manganese removed by upfront oxidation to particulate forms
- Hardness removed by lime-alum or lime-soda dosing
- TSS removed by coagulation/flocculation and settling sedimentation
- Fine polishing done by multimedia filtration
- Direct feeding of city water to demineralizers may not need any treatment, except for removal of chlorine and dissolved organic compounds by granular activated carbon (GAC)
- Treated water also used as make up for cooling water in closed loop
  - Continuous recirculation and cooling in cooling towers
  - Treatment includes disinfection by biocide and corrosion control
  - Continuous make-up and blowdown
  - Blowdown routed to effluent treatment plant (ETP) to mitigate any tube leakage through cooler/ condenser

Water Treatment Train

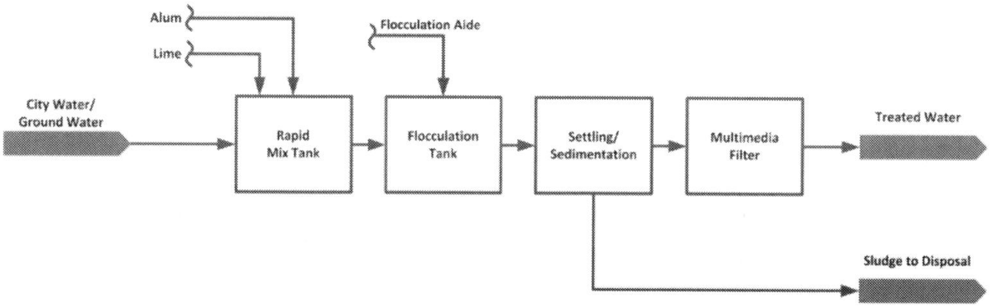

Water Demineralization

- Purpose of demineralization - remove inorganic (and ionic) TDS and prepare water for feeding boilers to raise high pressure steam
- Steam used to drive power turbines, vacuum ejectors and in strippers and distillation columns
- Treated water fed to demineralizer
- Demineralization process by IX, or RO, or their combination
- IX process consists of:
  - Cation removal
  - Decarbonation
  - Anion removal
  - Fine polishing for silica removal
- RO process consists of:
  - UF to remove suspended solids
  - RO to remove dissolved solids
- IX resins regenerated periodically by caustic and acid solutions that produce waste streams requiring pH neutralization before disposal
- RO membranes produce a fraction of the feed as concentrate that needs proper disposal

Water Demineralization and Condensate Polishing Trains

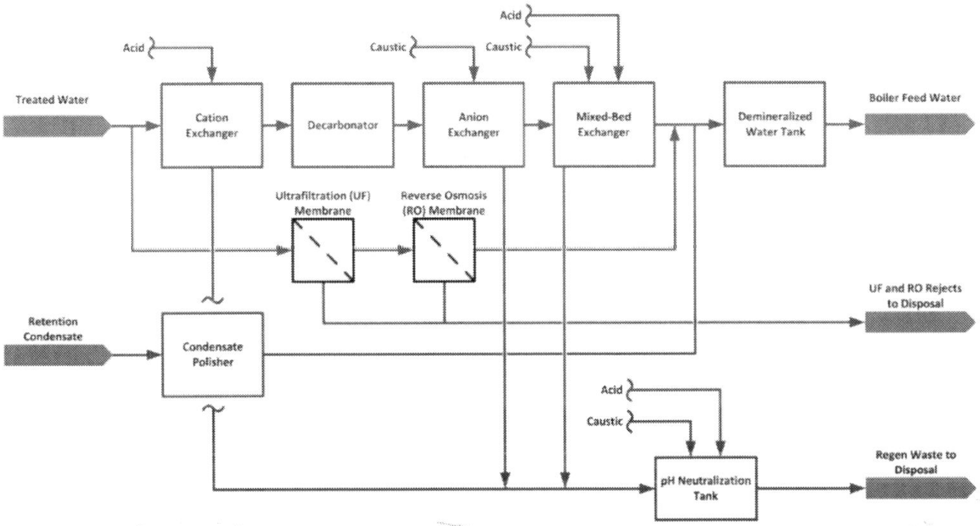

Boiler Feed Water (BFW)

- BFW has to be DO free to control corrosion of steel tubes in boilers at high temperature
- DM water further treated in two steps to remove DO
  - Deaeration by LP steam stripping
  - Further DO removal by internal chemical treatment, e.g. dosing sodium sulfite (Na_2SO_3)
- Chemicals added for internal treatment concentrate in boiler drum and continuously purged out as blowdown
- Boiler blowdown high temperature and concentrated in dissolved solids - need proper disposal

Condensate Polishing

- Condensate streams from non-process exchangers/ condensers, e.g. turbine surface condensers, heating steam condensates, collected steam trap condensates, etc.
- Condensate streams become contaminated by iron and other metals from direct contact with heat exchanger and condenser tube materials
- Metal ions removed from collected condensate stream in a polishing unit by ion exchange

Process Water Cycle

- Process water definition
  - Water/condensate streams containing hydrocarbon, H_2S, NH_3, etc.
- Sources
  - Condensate streams from process heaters /exchangers that can be potentially contaminated due to tube leakage
  - Aqueous streams separated from distillation column overhead
    - Crude Distilling Unit (CDU)
    - Vacuum Column
    - Other aqueous streams from stripping stream and vacuum ejector condensates
- Handling and Management
  - Sour water stripping
  - Desalting
- Fate of Process Water
  - Effluent treatment
  - Treated effluent disposal and/or recycle and reuse

Sour Water Stripper

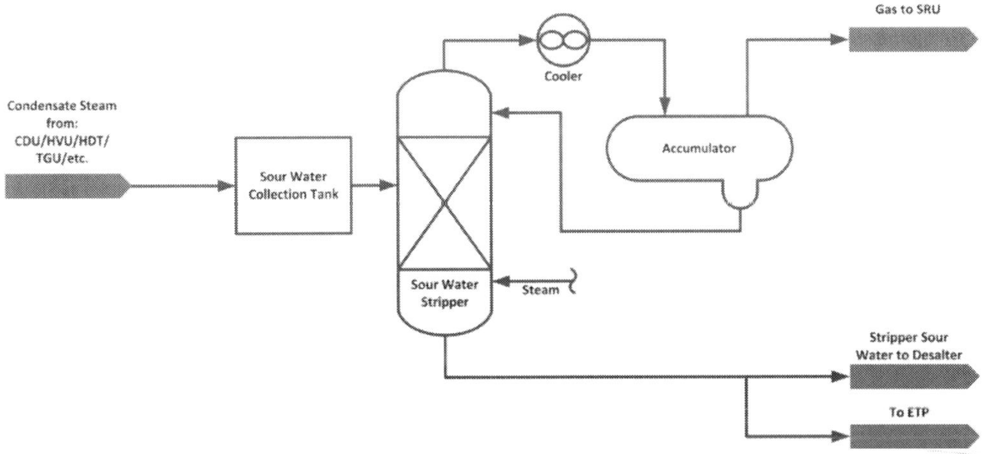

Sour Water Stripper (SWS)

- Process water streams contain H_2S, NH_3, low level of Cl^- and hydrocarbons
- These are HAP and need removal as these cause human health risk
- Removed in SWS
- Tray, or packed columns
- Liquid (sour water) fed from the top and flows down by gravity
- Stripping steam fed from the bottom
- Overhead vapor contains hydrocarbons, NH_3 and H_2S
- Stripped sour water (SSW) exits bottom of the stripper
- Typical product (stripped sour water) composition: $H_2S<50$ mg/L and $NH_3<100$ mg/L
- Single vs. double column
  - Single column removes both H_2S and NH_3
  - Double column – first remove H_2S at low pH, second removes NH_3 at high pH
- Performance depends on matrix – Cl^- and CN^- ions hinder ammonia removal; hardness, Fe and Mn cause scaling and fouling, free phase oil cause fouling

Desalter

- Desalter: first unit operation in a refinery
- Purpose: wash the crude with clean water (6% v/v)
  - Remove salt
  - Remove fine solids
  - Heat up to lower viscosity of crude and mix up with water thoroughly
  - separate water from crude and transfer it to waste stream - brine
- Crude contains:
  - Hydrocarbons
  - Formation water (connate water) – dissolved (<0.05% v/v) + dispersed (0.1 – 2% v/v, <10 μm)
  - Dissolved salts (primarily chlorides of sodium, calcium and magnesium) – 3 – 10%
  - Heterocyclic organic compounds containing N, S and metals (V, Ni, As)
  - Naphthenic acids (Total Acid No. – TAN, mg of KOH/gm of crude)
  - Amines used from production to control H_2S
  - Sediments/solids (clay, silica, iron oxide, calcium salts, etc.)
- Contaminants in crude and their impacts
  - Salt causes corrosion, fouling and catalyst poisoning
  - Hardness causes scaling
  - Water reduces capacity, increases energy consumption
  - Solids cause fouling and scaling

Desalter

- Feed – crude + washwater (~6% v/v)
- Products – desalted crude (<0.5% v/v water) + waste brine (<500 ppm oil, <500 ppm solids, <1000 ppm salt)
- Chemical dosing – Demulsifier, caustic, other additives
- Operating principle:
  - Dispersed water coalesced and separated from oil under electrical field at high voltage up to 30 kV
  - Demulsifier facilitates oil-water separation
  - Salt partitions with water and exits with brine
  - Solids settle into the water phase and exit with brine

Desalter

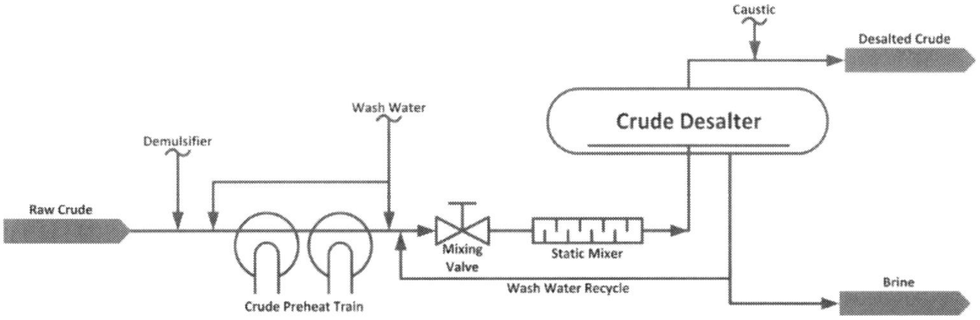

Desalter Vessel

Desalter Schematic View

Source: www.shakwmakw.com

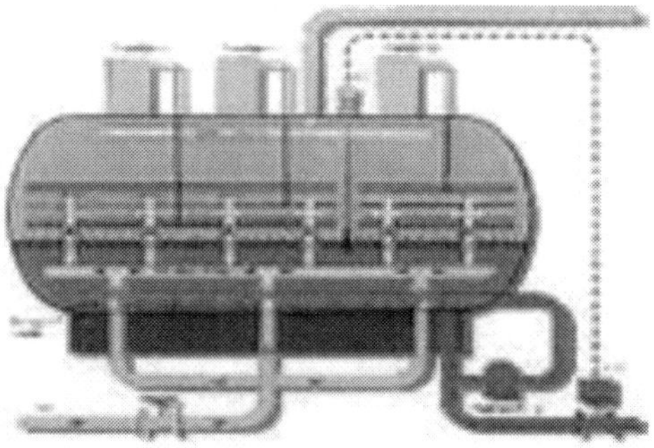

Desalter

- Operating conditions:
  - Temperature: 230 – 270 F
  - Pressure: Maintained >25 psi above total vapor pressure of crude and water to prevent boiling
  - Optimum Mixing: uniform mixing, no emulsion
  - Washwater: SWS effluent, service water, DM water
    - pH = 6-8, Cl^- <100 ppm, NH_3 (total) <150 ppm, H_2S <50 ppm, DO <50 ppb
- Performance specifications:
  - Cl^- removal efficiency: >90% single stg, > 97% dbl stg
  - Electrical spec.: <60% of dead short amps
- Operating issues:
  - Poor oil-water separation due to emulsion formation
  - Low desalting efficiency
    - High grid amperage
    - High chloride carryover causing column overhead fouling and corrosion
    - Caustic injection
    - Oil undercarry – effect on ETP
    - Water carryover – effect on distillation and energy consumption
    - High DO causing corrosion

Brine Disposal and Wastewater Treatment

Headworks INTERNATIONAL

- Desalter brine sent ETP prior to disposal
- Other streams into ETP
  - Excess stripped sourwater
  - Contaminated stormwater
  - Condensate streams from process heaters/exchangers
  - Spent caustic and other process waste streams
- Treatment steps
  - Segregation
  - Collection and equalization
    - Hydraulic, pH and chemical equalization
  - Primary
    - Removes solids and Oil & Grease
    - Reduces treatment load of secondary step
  - Secondary
    - Biological treatment to remove organic carbon and nitrogen
  - Tertiary (if needed)

Headworks INTERNATIONAL

Wastewater Treatment

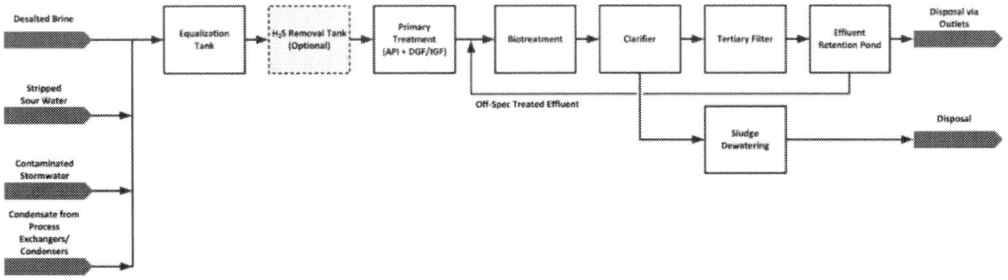

Fate of Treated Wastewater and Environmental Requirements for Discharge

- Achieve regulatory requirements of COD, BOD, TSS, NH_3-N, Oil & Grease, and Effluent Toxicity
- Typical NPDES permit limits for BOD and TSS <30 mg/L each
- Some locations impose more stringent limits, including TN
- Some locations may require TMDL compliance (Total Max Daily Load)
- Effluent discharge requirements determine treatment process
  - For discharge to surface water low TN and Effluent Toxicity required
    - Nitrification/denitrification for low TN although nutrient limits are not yet common to refineries.
    - Carbon adsorption by PAC or GAC may be required to achieve effluent low-level toxicity
    - Multimedia filter may be required to achieve low TSS consistently
  - For reuse, ultra low BOD/COD and TSS may be required
    - Will require treatment by a membrane bioreactor (MBR) process and/or activated carbon (GAC)

Regulatory Drivers

- Wastewater Treatment and Air Emissions Control Inseparable due to Hazardous Air Pollutants
  - Clean Air Act Amendments of 1990
  - Emissions from Point and Fugitive Sources – LDAR
  - Benzene NESHAPS
  - EPA, API and NFPA Standards and Guidelines
- Toxicity Monitoring and Reduction - WET

Summary and Conclusions

Headworks INTERNATIONAL

- Effective water management is critical to the success of uninterrupted petroleum processing operations – both upstream and downstream
- Desalter plays a very important role in petroleum refineries central to both hydrocarbon processing and effluent treatment
- Growing trend of use of opportunity crudes makes the downstream hydrocarbon and wastewater treatment processes vulnerable to upsets, leading to unscheduled shutdowns
- Such challenges can be overcome, but require study of each situation on a case-by-case basis
- Each problem is unique and there is no generic solution

Headworks INTERNATIONAL

Thank you

Thank you

Questions????

Headworks®

INTERNATIONAL

11000 Brittmoore Park Dr.
Houston, TX 77041

Phone: 713.647.6667
Fax: 713.647.0999
E-mail: hw@headworksintl.com

www.headworksinternational.com

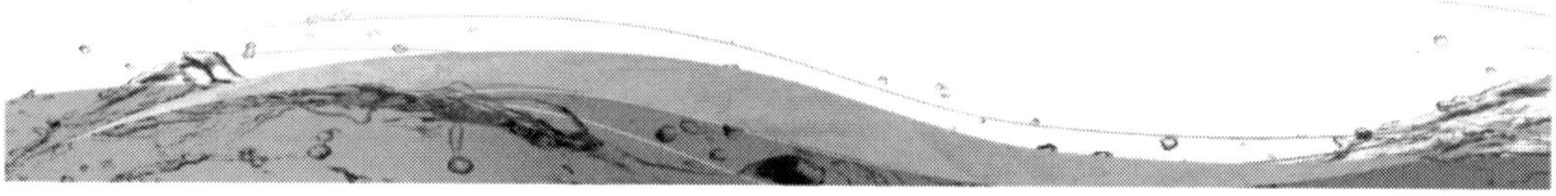

AUTHOR INDEX